SÁILLOONS ÁND FLIPTÁCKERS

This 1958 model Aero-hydrofoil embodied the essential features of the Fliptacker; all foils were designed for flow in one direction and tack was changed by flipping. Controlled by an air rudder, which became the lee hydrofoil on the opposite tack, it sailed very slowly, having too small an airfoil for the large buoyant hydrofoils.

SAILLOONS AND FLIPTACKERS

The Limits To High-Speed Sailing

By Bernard Smith

Drawings by Evan Mann

Published and distributed by

American Institute of Aeronautics and Astronautics

370 L'Enfant Promenade, S.W., Washington, D.C. 20024-2518

Contents

CHAPTER 1

Thirty Years of Change
in Review

To one who has turned lifeless materials into a thing alive and forced it to do his bidding against the resisting forces of nature, in silence, without fuel, and without defiling air or water, there can never be anything more wonderful than the sailboat.

To one who has not had the experience, no telling of it can touch him. The sailboat never offends the senses of fish, fowl or man; to make it move faster is to make it yet more a thing of freedom and beauty.

*T*hat reflection, drawn from my book *The 40-Knot Sailboat*, expresses my fascination with sailing as truly today as when it was written in 1963. I doubt it will change. However, the concept of the ultimate sailboat has indeed changed greatly from what was conceived a quarter of a century ago. For the first two decades of that span, my theoretical and experimental activities followed *40-Knot Sailboat* postulates that, while illuminating, turned out to be more nearly fixations that hampered further innovation.

Although many important advances ornamented those years, several critical deficiencies were never fully overcome. The oversimplified control system of the Monomaran was uncomfortably coupled to the lee hydrofoils, forcing some undesirable pitch behavior in high winds. The control simplification also slowed down the maneuver for changing tack. But the characteristic calling most for a changed approach was the burden laid upon sail and hydrofoils alike for reversal of flow, thereby placing

I have come to believe that, with few exceptions, modern attempts to break the sailboat barriers are mostly brute-force affairs precariously dependent upon ballast placed at the ends of enormous moment arms or upon bulging sponsons stuck onto the sides of squat monohulls.

undesirable limits upon efficiency achievable in those bodies. The work has been summarized in "New Approaches to Sailing," published in the March 1980 issue of *Astronautics & Aeronautics* magazine.

Sometime in 1982, a psychological maelstrom struck all the old ideas, forcing a full departure into new territory. Actually, this can be described more accurately as a complete oversight of a much earlier notion, toyed with in 1958. There, on page 59 of *The 40-Knot Sailboat*, reproduced here in the frontispiece, is a picture of it with wrong proportions and without benefit of essential theoretical understanding. In that version no component suffered the need for fore-and-aft reversibility, and thus could be designed in accordance with the best foil practice, with proper leading and trailing edges and fixed camber. Best of all, it had the potential for changing tack quickly by rapidly tilting the airfoil overhead from one side to the other.

After a 30-year hiatus I have returned to this early bias. The ultimate form is the "Sailloon," a craft partly supported by thick buoyant hydrofoils and partly by a flattened helium-filled blimp serving as a sail. Although it has the potential for the highest speed of all at any wind strength, I do not expect to see it assembled on a scale suitable for carrying a man — not in my time. The explanation is given in the chapter on the Sailloon, in preparation for the chapter covering the ''Fliptacker,'' a more moderate craft with fewer fabricating and operating problems, paid for by somewhat lower performance.

I have chosen not to spend here as much print commenting upon all the fast craft on the horizon as I did in *The 40-Knot Sailboat*. This may seem parochial; I will not contest the charge. But I have come to believe that, with few exceptions, modern attempts to break the sailboat barriers are mostly brute-force affairs precariously dependent upon ballast placed at the ends of enormous moment arms or upon bulging sponsons stuck onto the sides of squat monohulls. A typical example, ''Crossbow,'' consists of a gondola for the crew placed out over the water on a great sliding bar to counter the capsizing torque of clouds of sail. It derived the name from a fear of losing one of its crew overboard when critically balanced, in which event the rest would be catapulted over the mast into the water on the other side. I prefer more subtlety. The winged keel of America Cup vintage and the beautiful rigid sails on some catamarans are more to my liking, yet they merely gain incremental advantages while contributing little to the battle against overturning. A slight edge in speed may be of prime importance to a contestant in a sailing race. I am much more interested in seeking a step-function advance in speed — two, three and perhaps four times the speed of the wind, just as an iceboat can do — but without brinkmanship.

I am talking about elegance, about capturing the insight that simplifies and unifies all the trivia into a breakthrough formula — the $E = mc^2$ of sailing. Here is to be found one of the few remaining highly rewarding enterprises left to individuals who wish to innovate without the help of hundreds of technicians, millions of dollars, and many unpaid critics. One doesn't even need the ubiquitous computer. In this endeavor the designer can remain a complete person, sometimes a scientist, sometimes an artist and, if need be, sometimes a theologian — one can build upon his ideas and test them all by himself much as he would cultivate a garden. And with modern materials to shape and strengthen his devices, he can work wonders with hand tools.

So we proceed to the long-awaited sequel promised in *The 40-Knot Sailboat*. Readers who may not be conversant with the documents cited here or with general sailing practice may find the text more comprehensible after a session with Appendices A and C.

CHAPTER 2

Fig. 1   *Prototype-scaled Aerohydrofoil, forerunner of the man-carrying model, struck the best compromises.*

The Aerohydrofoil
and
the Monomaran

*T*he theories of high-speed sailing evolving in my mind were always tested with complete platforms; in part because of my belief that the principal sailboat components are inextricably interrelated and in part because I have been too impatient to await component tests in wind tunnels and model basins. Often I was far too busy with other responsibilities to do otherwise. As a result I often milked hasty conclusions from infrequent furtive launches on small bodies of water, inadequately instrumented for good data. It may be said that I practiced great economy in decision-making, employing an absolute minimum of facts.

The net result of such sporadic experimentation — including two abortive trials with full-scale versions, all spread over a period of ten years — was the radio-controlled model shown in F-l. Dubbed the Aerohydrofoil, it seemed to strike the best compromises. Its tightly stretched and cambered sail operated well enough at small angles of attack to give the desired performance when beating to windward at high speed. Inclined to windward at about 30° the fixed sail produced true lift and cancelled listing moments. Moreover, because of its stiffness, normal-force damping — rarely experienced with other sails — imparted to the model a steadying effect, equivalent to adding inertia.

The fixed lee hydrofoil was designed to operate equally on either tack and was curved to turn the force more vertically as it rose with increasing speed. On the other hand, the two articulating weather hydrofoils were curved to provide high lift when fully immersed at low speed and very little lift when raised to the proper depth of immersion. The weather hydrofoils gave directional control. The model was given no hull, in the hope of eliminating transitional problems in changing from hull support to foil support. Events were to prove otherwise. Initial support was supplied by the inherent buoyancy of the thick-rooted hydrofoils. When the craft came up to speed, about 10 kt, that support was replaced almost completely by the dynamic lift of airfoil and hydrofoils. This model changed tack by "wearing" about — by reversing its direction and adjusting its hydrofoils accordingly. In other words it was a "double-ender." It performed so well

Fig. 2 *Full-scale, man-carrying Aerohydrofoil moves upwind at 15 kt in a 10-kt wind. The platform has risen 3 ft from its rest position. Poor low-wind performance and sensitivity to weight led to its abandonment.*

A properly shaped displacement hull simply cannot be beat as a low-drag load carrier at low speeds. In fact, a well designed sailboat with a very good hull can generate speeds greater than the wind speed in winds that do not force it beyond its critical hull speed.

in rough and smooth waters that in 1969 it motivated my decision to construct a man-carrier along the same lines.

The man-carrier was spectacular. In winds of slightly under 10 kt it moved on near reaches at over 14 kt. In a wind of 12 kt on a similar point of sailing I estimate it reached 20 kt. This is believable performance since theory calls for just such a progression. At the higher speeds the platform lifted 3 ft above its rest level. At no time in its maneuvers was the slightest listing or pitching observed. In the accompanying enlargement from a motion picture of a 15-kt run (F-2), despite the poor photographic resolution you can make out the smooth wake left by the hydrofoils. The flow over the hydrofoils throughout the tested wind spectrum of 3 to 12 kt left the impression that the larger version was behaving much like the small model and that the transition from displacement lift was uniform owing to the elimination of the hull. More careful measurements taken at the low-wind end revealed that it was not so. The water could not be fooled.

In a wind of 8 kt the speed of the man-carrier was not over 7 kt and in a wind of 6 kt, not over 3. This craft was experiencing a displacement ''hump'' just as if it had a hull, something not exhibited by the model. I soon realized that the scale model had a considerably lower ''hump'' speed, not witnessed because under the previous test conditions it had invariably been exceeded. Unfortunately, low winds were the rule over the local waters and this craft was no better, if as good, as ordinary sailboats under such conditions.

Despite the fact that it lacked a proper hull, the Aerohydrofoil displayed much the same characteristics of other hydrofoil sailboats fitted with conventional hulls. In common with them, it was a poor sailer in low winds, where a sailboat is better off without the drag of submerged hydrofoils. Nevertheless, it had two distinct advantages over the others in strong winds, both derived from its asymmetrical arrangement. It could carry more sail and it could sail closer upwind.

Other idiosyncracies began to appear. In the shallow local waters the nearly 5-ft draft of the hydrofoils at rest was a serious inconvenience. But worst of all was the machine's sensitivity to payload. Any attempt to add another passenger to this 25-ft-long sprawling platform with 330-sq-ft sail would prevent any liftoff even in 12-kt wind. This may seem a bit startling until you realize that the addition of a 180-lb man to the original total load of 430 lb, including pilot, represents an increase of over 40%.

Better compromises were needed. Some speed could be sacrificed at the high-wind end, where there was plenty, to improve speed at the low-wind end, reduce draft, and improve load-carrying ability. To upgrade performance and load capacity in low winds meant reintroducing the hull. A properly shaped displacement hull simply cannot be beat as a low-drag load carrier at low speeds. In fact, a well designed sailboat with a very good hull can generate speeds greater than the wind speed in winds that do not force it beyond its critical hull speed. In retrospect, it can be said that the mistake made in this case was to design into the hull too many other functions. Making the slim hull both double-ended and asymmetrical, thus taking up the function of the replaced weather hydrofoils, somewhat increased drag. Articulating the hull ends for control produced additional flow interferences. The unique part of the new design was replacement of the lee hydrofoil with two swiveling water-skis, mounted at the lower extremities of the sail (F-3). These worked quite well, although they diverted the work in the wrong direction. I called the new machine the Monomaran, since it employed one hull to carry the load yet obtained its stability more in the fashion of the multihulled boats.

These changes reduced the draft to 18 in. for a three-man payload. Testing revealed that the low-wind performance was indeed improved but not really enough to compensate for the substantial loss

Fig. 3 Monomaran I had an asymmetrical hull and water-skis to support the airframe. Low-wind performance was improved but at an unacceptable loss of speed in high winds.

of speed at the high-wind end. Moreover, the machine was cumbersome and taxing to control, not at all like the lively Aerohydrofoil. This little project remained pretty much in limbo until 1975 when I was in retirement and received reinforcements in the form of Frank Delano, who had learned of the Aerohydrofoil and was anxious to see its development completed. Soon a scheme was laid out whereby the work could be resumed by cannibalizing the old components and "liberating" other sailboat parts at hand.

There seemed to be no escape from the mournful task of returning to a more conventionally shaped hull incorporating a centerboard and rudder. Yet the spell of the water-skis, which provided no leeway resistance or directional control, was still too strong. The stability derived from the sailframe and water-skis removed any need to impose roll-stability on the hull, which therefore could be designed primarily for low resistance. The simple process of reversing direction to change tack was abandoned. The hull itself was now steered to come about in the usual fashion of sailboats and, by means of an aerodynamic trick, the sailframe was forced to swing in front of the hull for the opposite tack — sometimes!

To understand the action, bear in mind that, in common with other airfoils, a sail's center of pres-

As for the airfoil, past experiences had shown that a tightly stretched, properly shaped sail was almost as good as a wing at low angles of attack. Faced by the obvious choice, the need for tightness was turned into a virtue instead of a problem.

Fig. 4 Monomaran II had a symmetrical hull and was maneuvered to come about like a conventional sailboat by employing aerodynamic forces to swing the airframe around the bow. Although an improvement over the first Monomaran, it had an unsatisfactory control response.

sure moves forward in the direction of the oncoming wind. Thus a "snubbing" line from the hull's bow to the sailframe's center could constrain the two parts in sailing orientation with the sail normally filled (F-4). When the craft turned sufficiently into the wind, however, the sail would backwind and reverse its force, although the forward shift of the center of pressure would remain unchanged. The reversed force drove the sailframe around the hull's bow until the same snubbing line constrained it on the other side. But the maneuver couldn't be carried out consistently. The inertia of the airframe was too low, which also added uncertainties to changing tack by turning downwind.

The arrangement was eventually discarded even though it delivered speeds nearly equal to the wind over the low-wind-speed spectrum. The design goals were contradictory. High speed required a high ratio of sail area to total mass, yet the control system demanded more, and self-defeating, inertia.

The crisis was overcome only by recognizing that our design thoughts were still too much those of water animals. An air animal would have sailed the airframe in its own right and simply taken the hull along as a tender. That idea led to a better compromise and deserves detailed description.

The air approach brought back the articulating hydrofoils employed on the abandoned Aerohydro-foil, but with notable differences. By mounting them on the leeward side to replace the water-skis, the moderately inclined hydrofoils were relieved of the payload-support function. This allowed reducing the initial submersion of the hydrofoils to more desirable values (about 18 in.) and eliminated any need to raise them dynamically for the purpose.

The frame holding the sail and the hydrofoils became self-contained and independent of any sailing aid from the hull, which was attached at just one windward point. The residual function of the hull was simply to carry the load and to be towed by the sailframe. This greatly simplified the design of the hull both structurally and hydrodynamically. In short, the Monomaran frame became an unmotorized power plant that employed the Aerohydrofoil to drive any suitable hull.

As for the airfoil, past experiences had shown that a tightly stretched, properly shaped sail was almost as good as a wing at low angles of attack. Faced by the obvious choice, the need for tightness was turned into a virtue instead of a problem. The sail was converted to a structural member and designed to replace three guy-wires in the standing rigging. Sheetlines, vangs, and all lines employed for controlling the stretch, twist, and position of ordinary sails were eliminated. In truth, the sail became a fixed air-keel, aside from being a constrained kite

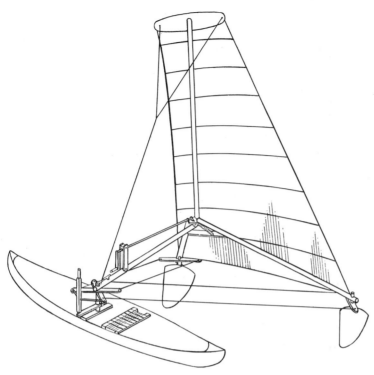

Fig. 5   Experiments culminated in a boat with fixed air-keel as "sail." The crew controlled it with hydrofoils.

(F-5). Only its inclination could be varied to reduce the driving force and listing moments as befitted sailing conditions.

With this arrangement, control of the craft devolved entirely upon the hydrofoils. A simple mechanism was improvised for turning the hydrofoils in parallel or in opposition. The first set of motions related to points of sailing and to speed adjustments. By turning them together one could obtain forward, zero, or reverse motions. To steer the craft left or right, one turned the foils in opposite directions. The control point was, of necessity, at the hull. The connections were made by means of cables running from the ends of spreaders on each foil to corresponding control spreaders at point of attachment to the hull. The control spreaders were linked in turn to a control stick that produced the desired hydrofoil motions in a very natural way: To move forward, backward, left, or right, the stick was tilted correspondingly.

Once more a problem was put to advantage. The control cables, which required substantial tension to operate properly, replaced the stretched guys holding the base of the sailframe together. In this way, member after member was integrated into the design until all duplicate components were eliminated. Even the crank that turned the frame-raising screw was made to pay out the windward restraining guy

Fig. 6 *Monomaran III, the final Monomaran, departs drastically from its predecessors. It successfully marries the original principles of the Aerohydrofoil to a substantial load-carrying hull. In this instance, the supporting frame itself is sailed independently, and the hull is towed as a freely trailing body.*

at the same rate, thereby keeping all the tension members automatically taut during raising, reefing, and lowering operations.

Originally the Monomaran's all-up weight less occupants was to be held to 140 lb and thus match the sail area in square feet. This matching relationship was later recognized as being applicable to the hull-less Aerohydrofoil but not essential for the Monomaran. The design simplifications brought the weight of the total assembly, including a 65-lb canoe, to 153-lb without passengers — considered close enough for the Monomaran although there was no doubt that the target weight could have been reached.

After the preliminary testing with a few hulls of differing design and correcting all symptoms of incipient weakness, a more extensive test program was begun using an 18-ft aluminum canoe (F-6). No operational or performance problems were encountered until winds and corresponding velocities reached 20 kt. By then, the canoe with two occupants was suffering from "stern squat" and being raised half out of the water by the sail's resultant. That was acceptable, but the loss of control when the forward hydrofoil also lifted out of the water was not.

The trouble was diagnosed as too much inclination of the hydrofoils. Not only did the sail's center

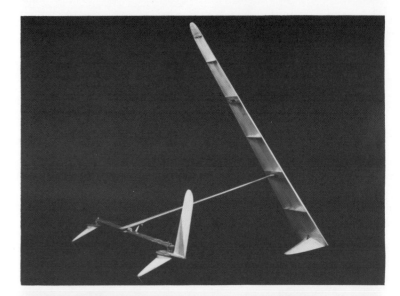

Some late 1950's thought in model form.

of dynamic pressure favor lifting the forward foil, but also the forward foil had inherently higher lift as a consequence of a directional requirement for higher angle of attack. Once the forward foil had lifted out of the water, control was shifted to the rear foil, which could not handle the job alone. In fact, any attempt to turn upwind with the rear foil alone caused it to be depressed in the water. This underwater sail was being "backwinded"!

Reducing hydrofoil inclination solved the problem. The sail's inclination was increased to make up the loss. It is illuminating to compare the area ratios (ratios of sail area to hydrofoil area) and wind strengths of the Monomaran and the Aerohydrofoil at 20 kt. The Monomaran barely reached a ratio of 50 and required a 20-kt wind. The Aerohydrofoil reached a ratio of 100 and needed a 12-kt wind. Nevertheless, in winds of less than 8 kt the ratios were closer and there the Monomaran could outpace the Aerohydrofoil.

In low winds, where rapid changes of tack could be needed to counter strong currents, the Monomaran was unfortunately woefully slow. No correction, short of paddling, was found. Furthermore, although the hull was unburdened from need for "double-ending," the sail and hydrofoils, not being so relieved, exhibited limited versatility.

Model Aerohydrofoil at top speed.

CHAPTER 3

The Sailloon

$\mathcal{J}$oin me now in a free-association exercise, wherein we shall imagine unusual combinations of known components, scanning them afterward for any reconciliation with reality as taught by our experience. In other words, we shall separate the flip parts of our brains, where talk is cheap, from the stern parts, where action is costly.

Imagine being intelligent air animals in a free balloon longing to do something more exciting than merely changing altitude. As we look down over a body of water it seems to be flowing beneath us. (Assuming the air to be at rest is perfectly valid and, for our purposes, more convenient than taking the water as reference.) How can we harness this apparent water-wind to pull us through the air? By simply tying to a sea-anchor with a water sail, a float fitted with a keel, and by learning to control its direction with extra lines. We also streamline the balloon to

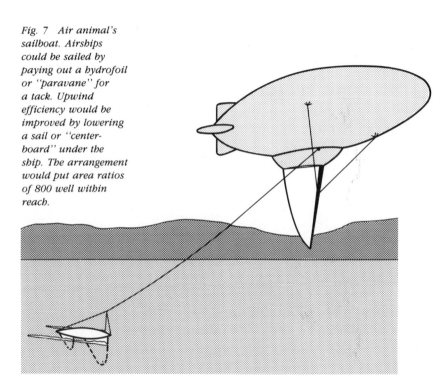

Fig. 7 Air animal's sailboat. Airships could be sailed by paying out a hydrofoil or "paravane" for a tack. Upwind efficiency would be improved by lowering a sail or "center-board" under the ship. The arrangement would put area ratios of 800 well within reach.

Fig. 8 A 100-ton blimp sailboat would have an airfoil towering 300 ft above the water. Two hydrofoils would be necessary for directional control. Payload space could be in the foot of the airfoil. When off-loaded, the craft would still need some weight in the hydrofoils; otherwise it would leave the water. The hydrofoils must therefore be proportionately larger than for the purely theoretical model, and accordingly the craft would be slower.

reduce its drag, and furnish it with control fins. Through trial and error we learn that, to be effective, the water-sail merely needs an area about 1/800th the projected area of what now is a blimp; and if we wish to retain contact with the water when off-loading, the water float should carry a substantial part of the payload.

The contraption resembles a friendly Goodyear blimp tethered in this case to a paravane, a controllable hydrofoil used by Navy helicopters to deploy mine-sweeping cables. This could be a pretty good sailboat — a better one if a true air-sail were broken out under the blimp as illustrated in F-7. In fact, if someone had pushed this idea in World War II, anti-submarine patrol blimps low on fuel could have tacked against strong headwinds and loitered near the shore until conditions allowed them to proceed to land destinations.

If a blimp was confined to ocean duties, why not convert its envelope into a real sail and attach it to a buoyant body near the surface, ending up with the rig illustrated in F-8? Here the flattened self-support-ing bag rests upon two large buoyant hydrofoils. This design is obviously the work of an air animal who thinks of supporting the payload with air dis-placement and employs hydrofoils to drive the bag through the air.

Because the sailing blimp is a true displacement vessel, it gains lift at a cubic rate and improves in stability as the scale goes up. A 100-ton craft of this kind could be held below a 30° list in a 25-kt cross-wind without ballasting. Since little of its surface encounters wave action, this cruiser would ride more comfortably than ordinary sailboats.

Owing to its large frontal area and relatively low inertia, such a craft could hardly be brought about on opposite tacks in the usual way. Positive control could be achieved by giving all the foils symmetri-cal, circular-arc sections and by wearing about. In this maneuver the craft would be reversed simply by turning the hydrofoils to provide a driving force in the opposite direction. A series of upwind tacks would leave a saw-toothed track in the water, unlike the sinuous tracks left by pleasure sailboats tacking to windward. The same strategy would be pursued on downwind tacks.

Because the relative wind would be so much greater than the true wind, angles of attack would tend to be low. At low angles of attack circular-arc foils compare well with the more common blunt-nosed airfoil profiles. This solution eliminates any need for reversing camber in the foils, a complica-tion that has always introduced many additional components and much unreliability, weight, and expense whenever it has been tried in connection with true wing profiles. The large shift forward of the airfoil's center of pressure works well for direc-tional stability. Since the smaller hydrofoils do not produce as much shift in their combined centers of pressure, the forward hydrofoil must take on a higher angle of attack to keep the craft on course. This automatically supplies a stabilizing dihedral angle between the hydrofoil chords.

Some critical characteristics of hydrofoils could modify the picture in high winds. Surface-piercing hydrofoils must keep to small angles of attack at high speed or they will experience "ventilation," a kind of stalling that drastically reduces the force and

Fig. 9 A blimp sail-boat model fitted with a gas-filled wing matched theoretical performance when dynamic forces raised it to the proper height. Lightly loaded in an 8-kt wind, the craft had "dry-to-wet" area ratio of 200, and made 20 kt on a beam reach. This model, tested in 1969, was perhaps the world's first Sailloon.

increases the force angle. To ensure that angles of attack are always sufficiently small, the area of the hydrofoil would have to be proportionately larger than called for by theory. The price is lower speed, but some additional buoyancy is obtained from the greater displacement. This can become important for another reason. Short of tethering or ballasting the hydrofoils with seawater or deflating the airfoil, the craft could not be off-loaded if too much support were derived from air buoyancy. Without a margin of water buoyancy it would leave the water.

I did put together a 9-ft model of a sailing blimp (shown underway in F-9). At this scale the airfoil lacked the buoyancy to lift the connecting structure, radio-control package, and other appurtenances, or even to add much vertical stability. Three hydrofoils supplied the necessary flotation and stability. True dynamic lift had to be built into the hydrofoils because at rest they had underwater areas far too great to yield the ratios called for in the idealized model. Additionally, all foils had to be inclined in such a manner as to cancel overturning moments because the model carried no ballast. In winds over 8 kt, however, the platform raised sufficiently to achieve the proper ratios and the little model beat to windward at better than the wind's speed. Nevertheless, the cost of the inflated wing and problems of gas leakage discouraged further experiments.

Fig. 10 A tanker could be sailed with the gases derived from the well yielding the oil. The objective of gas-bag design should be a shape yielding an L/D of 4 or better (as good as most sails). A 5-million-cu-ft container could drive an average tanker at clipper speeds. Owing to the lower-than-air density of wellhead gases, no rigid support would be needed. In changing tack the bag would swing overhead to the other side as the ship made its turn. The restraining cables would be winched as necessary during the maneuver.

The Sailloon is disarmingly simple in concept.
To begin with, nothing smaller than a costly 60-ft
wing could contain enough helium to support an
average man; the rest of the weight would have to
be born by the hydrofoils. Furthermore, at the minimum scale the righting forces would be quite low
and the operator would thus need to perform some
acrobatics on an outrigger to keep the platform
erect. And anything larger would take many millions
of dollars to solve all the problems of construction,
porting, launching, reefing and other operations, all
presenting formidable tasks without precedent.
Given this prospect, we must turn reluctantly from
this Sailloon design and look for better compromises.

The notion of driving a surface vessel with a
tethered blimp may have application to the trans-
portation of lighter-than-air gases such as methane
or ammonia. Well gases, mostly methane, today are
liquefied and shipped to destinations in cryogenic
pressure vessels. There the liquid is again gasified.
Why not fill a streamline bag with the gases in the
first place and use it as a sail to drive the tanker
transporting the oil? The bag could return deflated
with the empty tanker for a refill of both gas and oil
(F-10), eliminating the need for cryogenic treatment
at either end and half the fuel cost of the round trip.
The Sailloon would seem to be cheaper than conventional cryogenic shipment.

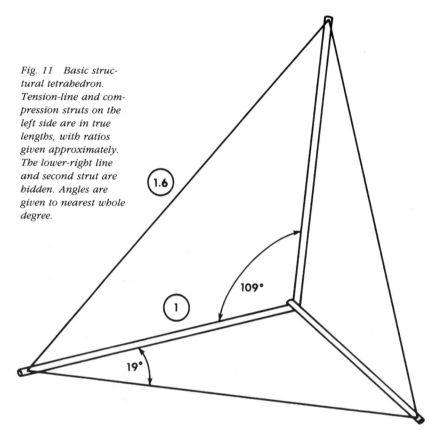

CHAPTER 4

Fig. 11 Basic structural tetrahedron. Tension-line and compression struts on the left side are in true lengths, with ratios given approximately. The lower-right line and second strut are hidden. Angles are given to nearest whole degree.

1.6

1

109°

19°

The Fliptacker

W e now depart speculation for more orderly action decisions, hopefully bypassing exercises in futility. First needing clarification are requirements for the "ultimate craft."

Moment Balance. The battle for lightness will go poorly if overturning moments must be countered by obesity — i.e., ballast, whether live or fixed. To slim the craft down, the force vectors representing the sail and its opposite underwater components should lie on the same line. Thereby the sail's moment arm vanishes and with it the moment as well.

Although this solution would ensure roll and draft stability (suitable for a Sailloon that met the ideal area-ratios) it would cancel the sail's lift component and with it the chance of reducing the wetted area of a craft overly endowed. The alternative is to resist leeway with vertical surfaces and counter the moment arising therefrom with a lifting hydrofoil placed as far to leeward as practical.

Coming about. An ultralight unballasted sailboat exhibits the antithesis of speed in changing tack. It has too little inertia, often gaining sternway before completing the turn. Correction must come in the form of reducing the angle needed to make the change and/or finding means quickly to bring the sail from one side to the other.

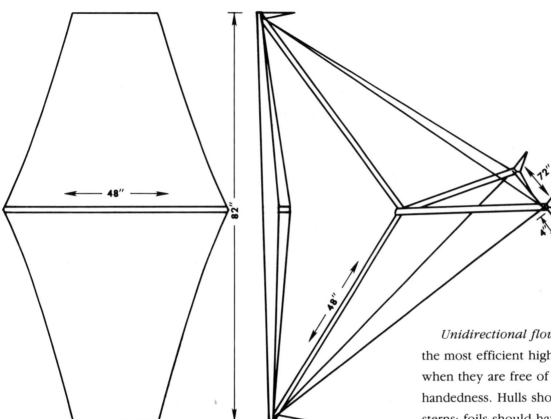

*Fig. 12   Proportions
of this Fliptacker
1/3-scale test model are
similar to those of a
potential man-carrier.
The space between the
windward hydrofoil
sets would normally
serve for positioning a
single hull, but no
orientation was found
entirely suitable for
both tacks. Moreover,
the single hull did not
lend itself to any
mechanism for auto-
matic flipping.*

Unidirectional flow.   Simple employment of
the most efficient high-speed shapes is possible only
when they are free of flow reversals or changes in
handedness. Hulls should have honest bows and
sterns; foils should have one leading edge and one
low-pressure side.

Connecting structures.   The frame most suited
to integrating these requirements seems to be the
tetrahedron. The version with the best strength-
to-weight ratio contains four internal compression
members tied together by six tension members at
the edges of the tetrahedron.

In the elemental tetrahedron (F-ll) the lower right
corner is identified as the windward edge, which
in this view is foreshortened to a point. The opposite

edge shows in true length, as do the two struts intersecting its ends. Each edge is roughly 8/5ths of the strut length and 35° from vertical. Struts are at 109° to each other and 19° from horizontal except for one that is vertical.

The next view, F-12, adds a bit of perspective and depicts a sail projected first in plan and then moved over to replace an edge of the tetrahedron. Since either end may contact the water at one time or another, a lifting hydrofoil goes on each. Hydrofoil sets are emplaced at the ends of the windward edge. Note that the model is symmetrical about a plane passing through the windward edge and the middle of the sail. To change tack, the model must swing through nearly 110° about the windward edge.

To ensure the verticality of one hydrofoil in each pair, as pictured in F-12, the members must be separated by a corresponding angle of 110°. In this form the model has no payload, directional controls, or means of varying parameters; but it will sail quite well if everything is just right. What does that mean?

It means first of all that the chords of the sail and the vertical hydrofoils must diverge at more than some critical angle (15°, vertex forward, would be enough for well-faired components). If, for example, the designer wants minimum drag on the sail when it flips overhead through a change in tack

Fig. 13 Fliptacker model as constructed. Shown first is the overhead position transiently passed by the Fliptacker when changing tack. In the second view the model is simulating a tack toward the viewer, as indicated by the orientation of the small hull to the left. The wind is from the viewer and may be as close as 15 deg from the left and still power the tack.

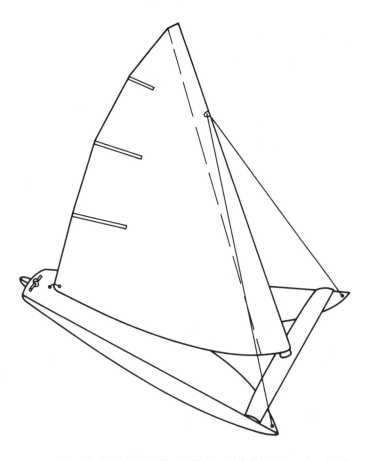

Fig. 14   Split-hull sailboat. This model did not require sheeting
except for running before the wind, where its performance was
poor. A list of approximately 15 deg raised the windward hull
clear of the water, giving the leeward hull the dominant reaction
to the sail's resultant. Turning approximately 10 deg across the
wind reversed the sail's camber, changing the list and the tack.
In a full-scale craft, the pilot would move around the mast to
remain in phase with the tack. Theoretically, direction could be
changed by shifting the pilot's weight — forward for upwind
and backward for downwind — thereby eliminating the need for
a rudder. If the lower hull edges were shod with metal runners,
this craft could negotiate ice.

(important for this light platform when turning
across the wind), the sail's chord should be parallel
to the windward edge and the vertical hydrofoils
should head 15° off the same edge. On the other
hand, to replace the windward edge with a hull,
the chords of the hydrofoils should be lined up
with it, otherwise the hull would be crabbing side-
wise at 15°. But this alternative forces the sail to
pass through a 15° negative angle of attack and adds
30° more turn every time the craft comes about.

I found the first solution better even though it
demanded the injection of an odd hull. More about
that later.

F-13 shows photos of the model that proved
very fast and controllable by virtue of ruddering the
aft hydrofoils. The last feature took care of the
changing drag relationships between leeward and
windward foils as wind strengths and craft speeds
changed. Alas, all this good behavior notwithstand-
ing, under no conditions would this model flip to
the other tack. Nor could any satisfactory orientation for a load-carrying component be found as long
as thinking locked on the windward hydrofoils.

It might have worked out as a Sailloon on a
larger scale, wherein buoyant lift would raise the
sail overhead as side force diminished in the turning
operation; but that possibility was droipped for
reasons given earlier. Not having that help, the model

simply dug its heels in when ruddered to turn, settling deeper in the water and actually listing in the wrong direction.

Consider a solution that works without a buoyant sail yet could work to advantage with one.

Split a husky surfboard lengthwise, exchange sides, and join them at their sterns to form a 30° "V," open end forward. Ballast this platform such that it can ride on one or the other hull, as the pilot wishes, by tilting its mate up out of the waves. Then sheet a fairly flat sail on a line that splits the angle in two. This combination yields a sailboat that can come about by merely turning from a course that just fills the sail on one tack to one that just fills it on the other (F-14). It works; but it needs a very nimble pilot; and I could never find one small enough to run the 4-ft models that I tested. Consequently, without live ballast, every version capsized as soon as it got going well (relative wind went way up producing an unbalanced moment).

What, if anything, can be gained by marrying a contraption that flips the wrong way with a Fliptacker that won't flip at all?

Recall that when one hull (F-14) is waterborne the craft moves in the direction of that keel. The other is *over* the water and oriented to glide over oncoming waves. Each hull now resists leeway in one direction only. We jack up the tetrahedron,

What, if anything, can be gained by marrying a contraption that flips the wrong way with a Fliptacker that won't flip at all?

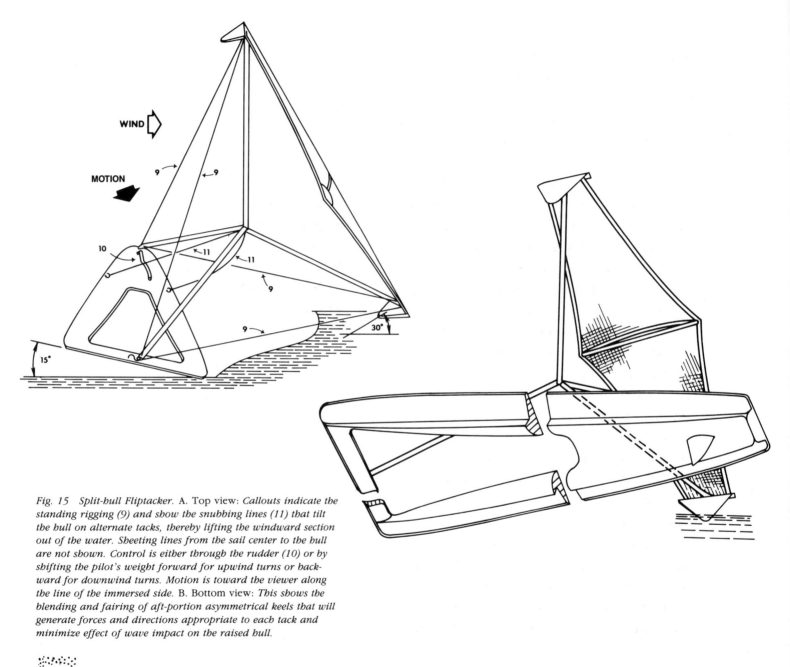

WIND

MOTION

15°

30°

Fig. 15 Split-hull Fliptacker. A. Top view: *Callouts indicate the standing rigging (9) and show the snubbing lines (11) that tilt the hull on alternate tacks, thereby lifting the windward section out of the water. Sheeting lines from the sail center to the hull are not shown. Control is either through the rudder (10) or by shifting the pilot's weight forward for upwind turns or backward for downwind turns. Motion is toward the viewer along the line of the immersed side. B. Bottom view: This shows the blending and fairing of aft-portion asymmetrical keels that will generate forces and directions appropriate to each tack and minimize effect of wave impact on the raised hull.*

remove the windward hydrofoils, and replace them with this way-out surfboard, attached along its midline. Where the boards join we put a rudder; and to make sure that the last bit of sail rotation lifts the windward halfboard out of the water, we tie a snubbing line to it from the sail frame (F-15). The succession of photos in F-16 shows the arrangement and the action.

This craft will flip if the trim and balance of the assembly requires a moderate side-wind to push the lee hydrofoil into the water and raise the drip-dry halfboard out of it. In that case a similar wind in the other direction will flip the sail over if the lee hydrofoil does not hook the water. (More about that later.) This craft, in short, goes far to meet the requirements listed earlier.

To be candid, the Fliptacker is not perfect. Every machine known to man has its limitations. What they may be for the Fliptacker will not be known until a man-carrier has gone through a great variety of environmental experiments. Due caution should attend expending time, materials, and manpower in a search for the perfection that is implied by the term "pilot-proof."

My point is made by the sailboard, the simplest high-speed sailboat ever invented, operable in all manner of waters by a skilled sailor. He senses the strength of gusts and waves in advance and reacts to

Every machine known to man has its limitations. What they may be for the Fliptacker will not be known until a man-carrier has gone through a great variety of environmental experiments.

Fig. 16   Split-hull Fliptacker photos show the stages of flipping from one tack to the other. The wind is from the viewer and, if freed, the model would move toward the viewer in the direction of the wetted hull.

them with conditioned reflexes. "Improved" into a machine that forgave all of the pilot's mistakes, the sailboard would lose its beautiful simplicity and become a complicated monstrosity that no doubt often would fail at the worst possible time (just like the complicated modern American automobile that cannot reconcile perfect reliability with perfect complexity).

I have described a simple Fliptacker, suitable for classification as a super sailboard, when operated by a skilled, resourceful pilot at one with his machine. When conditions arise in which moments do not exactly balance, he accepts a small list or corrects with his own weight. When the wind is not strong enough to flip the sail or the lee hydrofoil digs in before flipping, he interferes directly with his own muscles on the snubbing line. (On that point I go into more design detail in the next chapter.)

For a scale larger than this Cadillac of sailboards, we can only speculate what could go awry and on possible solutions. Probably the worst situation to contemplate would see the lee hydrofoil plowed under by a backwinded sail. The foil is now developing great hydrodynamic forces in the downward direction that keep increasing as its immersion increases. Under such conditions the sail cannot flip of its own accord and, if the platform is large, manual strength cannot cope with it. The simplest solution is a hinged hydrofoil that will swing free when the force is reversed. Next in severity could be a sail and structure too large and heavy to be pulled over when the wind is too light. That case may require more complication, such as a set of ailerons controlled by a joystick.

CHAPTER 5

The Super Boardsailer

When introduced a few decades ago, the sailboard's promise of cheapness, lightness and simplicity of operation quickly captured the imagination of the small-craft sailing crowd. Obviously, no man-carrying sailboat could be smaller, and certainly nothing could be simpler than using the man as a mast or steering a rudderless craft by shifting the sail's center of pressure. Indeed, if the pilot did not shrink from an occasional dunking, the promise was met by reality.

At first the potential of the sailing surfboard for record-breaking speed was not stressed. That was discovered later by the more daring operators, who noted an increase in speed when leaning to windward with the sail, almost to the point of overturning. A drop in wind strength could of course leave the pilot unsupported and deposited in the water; but that was all part of this great wet-sport, requir-

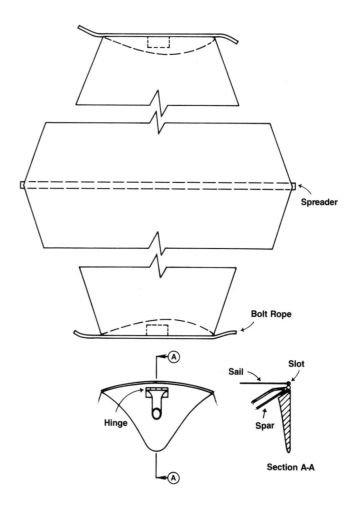

Fig. 17 Sail and hydrofoils of a minimum-size manned Flip-tacker. Hydrofoils are held in position by sail tension. The location of the hinge holding the hydrofoil to the spar is critical. It must be centered between the crown and the edges of the sail. In normal operation the hydrofoil angle will be limited by the constraint of the spar. When hydrodynamic force is reversed, however, the sail resiliency allows the hydrofoil to yield and relieve the pressure.

ing far more skill and talent than plain surfboarding or waterskiing, and supplying far more freedom and thrills.

Any observer conversant with vectorial mechanics could readily account for the substantial increase in a sailboard's speed when operated with the sail inclined to windward. In this configuration the sail generates a true lift component that can reduce the board's wetted area if the underwater drift-resisting components are not at the same angle. Unlike ordinary sailboats, the sailboard's underwater fins are not coupled to the sail. In fact, the clever operator can regulate the board's inclination with his feet to generate a similar upward force in the fins, thereby adding greatly to the total effect.

In short, the sailing surfboard piloted by a fearless acrobat can do with live ballast all that most dynamic platforms described here are designed to do — excepting only the various Sailloons. But suppose one wishes to remain fairly dry, or to exercise less acrobatic skill or to be independent of the live ballast or any shifting ballast whatsoever, or to graduate into a larger scale. What then? That is what this book is all about: not to replace the sailboard (Heaven's, no!) but to move on to the next stages. Moreover, just as the operators have done with the sailboard, the operators of the full-scale Sailloons

and Fliptackers will discover sailing potentials never imagined by the designer.

The time has come to talk of a minimum manned Fliptacker, and that will mean a sailboat designed around a surfboard or something close to it. Already discussed have been the general characteristics of the flipping sail assembly consisting of a wing carrying hydrofoils at its tips with four spars and connecting cables tying the whole into a tetrahedron. The hull lends itself to separate description since it joins the above at only two points, between which it can take many forms.

Particulars for the sail and the hydrofoils are given in F-17. The ends may be cemented directly to the hydrofoils or furnished with 1/4-in. bolt ropes for insertion into slots on the hydrofoils. The spars are standard 12-ft lengths of 6061 T6 aluminum tubing 1.5 in. in outside diameter and 0.05 in. in wall thickness. If the sail is held to the indicated span of 20.5 ft, then the four 1/8-in. stainless-steel cables should be 19 ft 9 in. between points of attachment. This leaves an 18-ft separation for hull-attachment fittings when all tension components are stressed. Fittings for spar joints and cable connectors should preferably be made of stainless steel. The central connector for all of the four spars is illustrated in F-18.

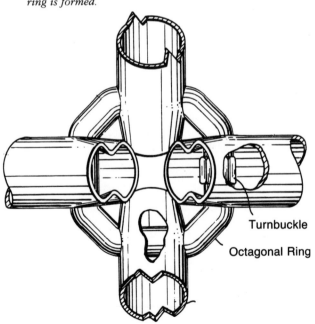

Fig. 18 For a central connector, an octagonal ring proves the simplest and lightest means of holding the four spars together while permitting the essential motion. Left- and right-hand threads must be cut on the ends before the ring is formed.

Turnbuckle

Octagonal Ring

Lightness is of paramount importance to the flipping assembly. Advantage should be taken of the newer high-strength, low-weight fabrics for sail construction. A Kevlar sheet would be ideal. Insofar as the hydrofoils are concerned, I have never found any materials more practical than styrofoam covered with glass cloth and epoxy resin. The foam is cheap and easily carved to shape, and when layered over with cemented glass cloth it becomes a very light, waterproof skin-stressed structure, unsinkable even when punctured. This system also works well for the hull.

To end up with a truly car-transportable disassembly, all components must be well under the dimensions of a small private car. The flipping subassembly presents no problem since it will knock down to a narrow bundle of spars 12 ft long. The hull is another matter; it forms a triangle, 18 ft on two sides and 9 ft on the forward end. It must be broken down into at least three components. When assembled it must be sufficiently rigid to take the tension to which it will be subjected plus stresses introduced by the pilot's actions as well as water forces, beach handling, etc. F-19 shows a suggested design. F-20 shows a model designed and built along these lines to be collapsible.

Raising of the flipping component proceeds in the following way:

1) Sails and spars are arranged on the ground in accordance with F-21.

2) The hull's stern is attached to the after spar with a short line, and the winch cable passing through the forward fairlead is attached to the forward spar.

3) As the cable is drawn up by the winch, the central four-spar joint and the uppermost sail spar are lifted manually until the winched cable alone can support them. (This will take an extra pair of hands initially.)

4) Continued winching then brings the assembly into tetrahedral conformity. The forward spar should be in contact with the hull's transverse spreader.

How can this sprawling, spidery behemoth be best baptized by partial immersion? The most practical first launching (mistakes most easily forgiven) would be undertaken from a gradually inclined beach in an onshore breeze. If correctly constructed, the entire assembly should weigh less than 75 lb and consequently should not be beyond the capacity of one man to shove off. With the sail to leeward, just enough to fill, the pilot should jump on as soon as the rudder can be lowered and turn the craft more offwind until the lee hydrofoil meets the surface. By then the craft should be riding solely

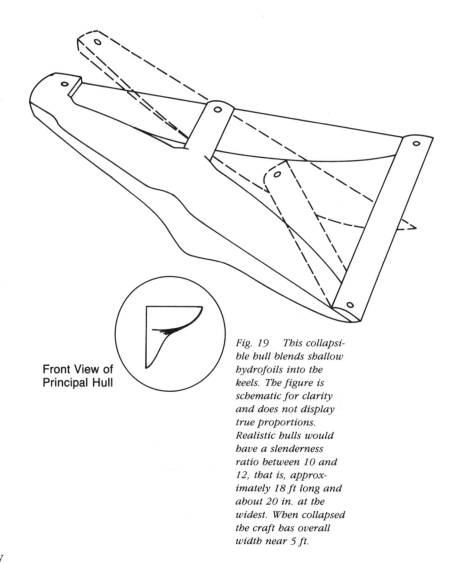

Front View of
Principal Hull

Fig. 19   This collapsible hull blends shallow hydrofoils into the keels. The figure is schematic for clarity and does not display true proportions. Realistic hulls would have a slenderness ratio between 10 and 12, that is, approximately 18 ft long and about 20 in. at the widest. When collapsed the craft has overall width near 5 ft.

Fig. 20   Fliptacker
model as designed
with collapsible hulls.
The sail overhead
shows the transition
position. The other
view illustrates a tak-
ing configuration:
windward hull lifted
by snubbing line as
sail foil contacts the
water. The keels can
be seen under the
hulls. The craft's mo-
tion follows the line of
the immersed hull
(with wind from
viewer).

on the lee hull; if not, the pilot should shift his weight until it does.

Once clear of shallow water, coming about is best practiced as quickly as possible. The craft should be gradually turned into the wind and at the moment of crossing, when the lee hydrofoil clears the water, the turn should be speeded up until the full wind flips the sail to the opposite tack. Should the wind not be strong enough, the pilot should aid the flipping by shifting his weight and pulling on the snubbing line before losing headway. In the event the maneuver cannot be completed before the lee hydrofoil submarines, its hinged joint can save the day. The pilot must then exercise his own live ballast as far as he can with the understanding that the hydrofoil will soon swivel the other way and he will be able to overpower its downward force. In that case he must be prepared for a sudden flip and brace for a quick reverse shift in weight.

Should worse come to worse, the pilot in dire straights, despite all efforts, need only release the dog on the winch ratchet and all will instantly collapse to the surface of the water. Hopefully he will have taken a paddle along for the ride back.

A Sailloon wing commensurate in size would lift about 15 lb, just about enough to ensure that the wing would indeed swing overhead when changing tacks and eliminate all the above apprehension.

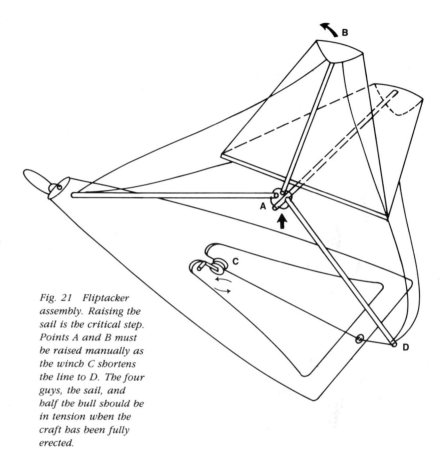

Fig. 21 Fliptacker assembly. Raising the sail is the critical step. Points A and B must be raised manually as the winch C shortens the line to D. The four guys, the sail, and half the hull should be in tension when the craft has been fully erected.

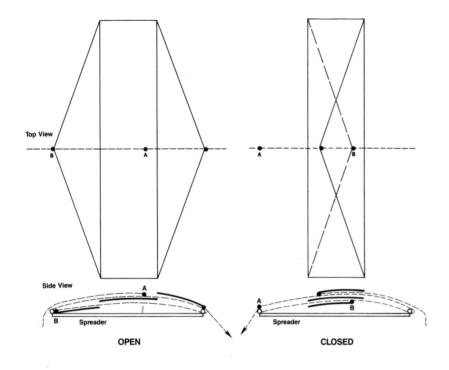

Top View

B   A

A   B

Side View

A

B   Spreader

Spreader

A

B

OPEN

CLOSED

Fig. 22 Variable-area sail. The positions of points A and B indicate the motion of the reefing lines in changing from maximum to minimum area. In practice, the lines on each side would be continuous, joining along the midline of the hull after passing through fore and aft pulleys.

Be that as it may, something still needs to be said about adjustments to wind strengths. F-22 suggests how the sail's area can be halved or alternatively made more effective with two slots. As indicated in F-22, a single "reefing" line accomplishes both ends. F-23 shows a sail designed to retain tension while changing area. The sail as expanded incorporates *two* slots — optimal for best L/D.

Finally, the frosting on the cake will come for the pilot when he recognizes that his speed made good downwind and will be greatest when he veers and tacks downwind, just as an iceboat would do, provided of course there is enough sea room.

How is this "Super Boardsailer" superior to an ordinary boardsailer? It requires considerably less acrobatics despite its far greater sail area — about 150% more than the best boardsailers.

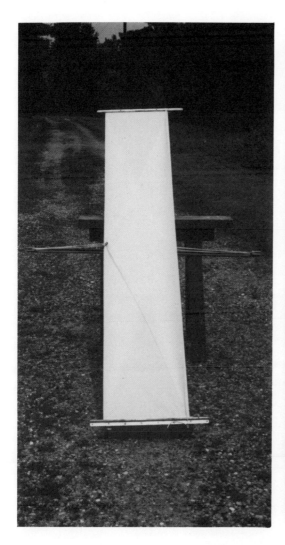
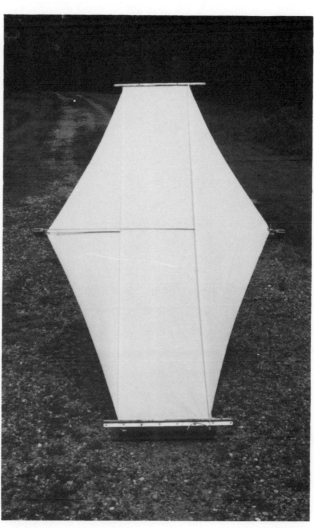

Fig. 23   Sail designed to retain tension while changing area. (The reefing lines are not visible.) The system follows the design of F-22 except the lines are returned to a central position on the spreader before passing over pulleys to the pilot. The change avoids sheeting the sail when reefing. The sail as expanded incorporates two slots— optimal for best L/D.

CHAPTER 6

Flipping Sailloon

*J*ust in case we find the angel who cannot resist financing the ultimate sailboat without thought of profitability, we shall hold in readiness a few of its particulars. Backtrack to the hull-less Flip-tacker and replace the single sheet with a helium-inflated wing opposed by underwater foils in parallel. Not needing dynamic lift, this model will have parallel spans of airfoil and fore and aft hydrofoils.

Given static balance through air and water displacement, the area ratio will be satisfied at all speeds. Moreover, by virtue of 300-lb lift of the inflated wing applied at a considerable moment arm, better pitch stability will compensate to some extent for the small water displacement at the extremities. However, that comes with a price. Because the same force resists roll, the sail's tilt angle is subject to the wind until its strength is enough to dip the leeward hydrofoil into the water.

About 6000 cu ft of helium will be needed to lift an average man and his mechanical support. A wing with sufficient volume would have these dimensions: span, 70 ft; midchord, 40 ft; endchord, 15 ft; thickness, 25% of chord. Being in spanwise tension, the wing will need ribs or partitions only in the middle plus ends to keep its shape, provided internal overpressure is much less than an ounce per square inch.

For the resulting wing area of 1700 sq ft, theory calls for about 1 sq ft for each immersed half of the windward hydrofoils. These will be roughly triangular, 1.5 ft on a side and thick enough to displace 6 lb of water each. More displacement would reduce the criticality of the wing's lift, but would also mean moving away from the ideal area ratio. The pilot will draw small comfort from the fact that complete immersion of both windward hydrofoils displaces only 24 lb of water. The craft would need to take on about 3 cu ft of water before the pilot could risk stepping off. Clever tank floats therefore need to be located in the vicinity of the hydrofoils. They are needed for yet another reason: the possibility that the Sailloon might lose its helium, in which case empty floats could support an operable craft; for even when deflated the wing is a pretty good sail, capable of driving this craft despite the float's added drag.

Tension trusses and a nacelle, in which the pilot sits and which can rotate freely by virtue of end bearings, bridge the distance between hydrofoils (about 60 ft). The nacelle remains upright owing to a low center of gravity, aided by the pilot's weight, regardless of how the wing frame flips. It offers little resistance to air flow because it is streamlined and normally aligned close to apparent wind (F-24).

From the pilot's position, flexible hydraulic lines to the rear foils give him control over direction of motion. He also manipulates valves from a helium flask to the wing to adjust its internal pressure and to purge water from the tank floats in accordance with the craft's load.

As necessary, the designer will specify exotic materials without regard to cost: Kevlar for the wing, laminated carbon filaments for the streamline struts and hydrofoils, titanium for the fittings, supersteels for the cables, etc. Total weight should be under 100 lb. Implemented in this manner, the craft will be a thing of grace and beauty and it should break the world's sailing speed record in a wind of about 12 kt (just enough to raise a few whitecaps). Much is left to be done — by others.

Caution dictates a few disclaimers. This ultimate sailboat is no paragon of convenience or economy.

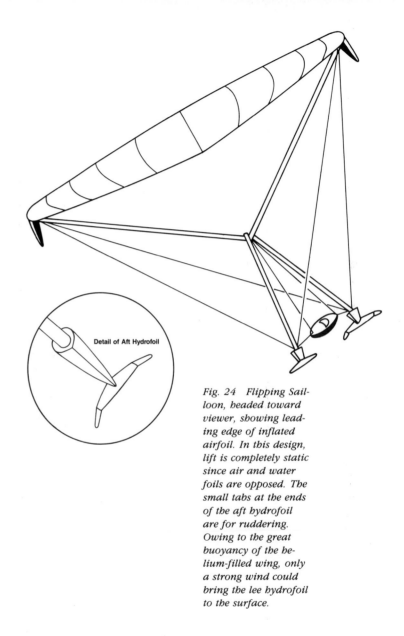

Detail of Aft Hydrofoil

Fig. 24 Flipping Sail-loon, headed toward viewer, showing lead-ing edge of inflated airfoil. In this design, lift is completely static since air and water foils are opposed. The small tabs at the ends of the aft hydrofoil are for ruddering. Owing to the great buoyancy of the he-lium-filled wing, only a strong wind could bring the lee hydrofoil to the surface.

Flipping Sailloons in general would lend themselves well to mooring or tethering. If attached at the right point the sail would remain overhead over a wide range of wind strengths — at times pulling in kite fashion to lift the craft clear of the water.

Aside from the ever-present specter of losing the craft to the sky if dumped overboard, the pilot would be kept busier than the proverbial one-arm paperhanger: eyeing the waterline, regulating gas pressure as temperature changed, purging or admitting water in the tank floats, judging the wind for a sailing course around obstacles or shallow water — all taking place presumably at high speed. Especially when testing an unfamiliar new platform, such busywork is a formula for disaster.

A training vehicle could relieve this situation to great extent, one that somewhat compromises speed for simplification, yet remains superior to conventional sailboats. The first step would be to cut air buoyancy to about 100 lb and to raise water buoyancy to 200 lb. If total unloaded weight exceeds 100 lb (say, 110) the Sailloon need never completely free itself from the water surface. Owing to the much smaller gas volume, far less attention would need to be given to temperature changes. Tank floats could then be replaced with honest hydrofoils, devoid of plumbing, displacing a little over 3 cu ft of water when fully immersed, thereby supporting about 200 lb of pilot and appurtenances. To regain dynamic lift, the windward foils of F-12 would be employed. With these changes the pilot can concentrate on the important business of steering to suit course and wind conditions.

The reduction in sail displacement cuts the craft's span to about 45 ft; and if proportions are preserved, all other dimensions will be cut correspondingly, bringing sail area down to about 900 sq ft. In low winds the wetted area would be approximately 18 sq ft, making the area ratio about 50. As winds increase, dynamic lift will raise this; at winds over 12 kt the area ratio would approach 500, not far from the ideal. In a 15-kt wind performance should come very close to the ultimate Sailloon's.

The sail's very low static lift introduces a problem: The pilot's nacelle may drag in the water at wind strength near zero. But there is some compensation residing in the greater ease with which the wind can tilt the sail to one side or the other. The much greater stiffness of the ultimate Sailloon would present exactly the opposite situation: lots of clearance for the nacelle, very little tilt in low winds.

Flipping Sailloons in general would lend themselves well to mooring or tethering. If attached at the right point the sail would remain overhead over a wide range of wind strengths — at times pulling in kite fashion to lift the craft clear of the water. Docking or beaching would not be as straightforward.

We now face the last problem: How do we get the pilot into his nacelle?

An early test of an Aerohydrofoil (June Lake, the Sierra, 1959), an imperfect design whose defects stimulated better concepts.

Theory of Sailing Limits

S ailing involves no thermodynamic cycle and generates little heat. Sailboats react mechanically to the forces of the wind without any train of energy-losing conversions in the path of action. As a consequence, the theoretical efficiency of transferring the momentum of a moving column of air to the momentum of a boat can be as high as the best windmills, even before windmills perform useful work: close to 60%. Sailing has the highest potential of any means for exploiting the cheap, renewable, clean power of the wind.

Sailboats and iceboats do not ordinarily climb heights to reach a destination; they simply move over an equipotential surface. Very light winds can set them in motion. However, as is abundantly clear from the superior performance of iceboats, the drag of an immersed hull severely limits the speed of sailboats. Acceleration and deceleration are of minor importance to sailing, coming into play primarily in

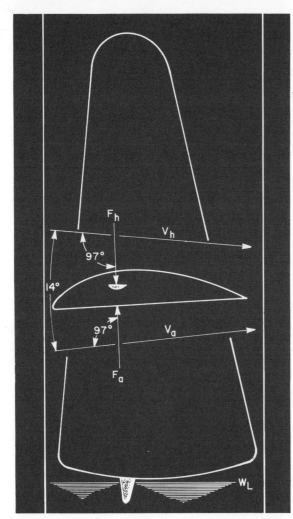

Fig. 25 The theoretical model. To attain equal speeds in air and water, the airfoil has 800 times the area of the hydrofoil. The inset shows angular orientation of foil sections for tacking. The force angle, 97 deg, corresponds to an L/D of 8. The forces developed by the airfoil and hydrofoil (respectively, F_a and F_n) are in equilibrum when air and water velocity vectors, V_a and V_n, converge at 14 deg. These characteristics give the model a potential speed four times that of the wind in a crosswind. The model does not address stability or control.

racing, where a rapid response to light gusts or an ability to change tack quickly can yield small but decisive gains, although a boat with insufficient inertia may not be able to come about readily.

Until modern times, sailboats were designed, understandably, from the viewpoint of a surface animal, namely, man. The problem was always perceived as one of harnessing the wind to drive a vehicle over the water with minimum danger of overturning. As a result, we have come to think of the wind as an upsetting force driving us off our course unless we supply ballasting and leeway resistance. On the other hand, sails have been articulated to "catch" the wind as it changes direction and strength.

We could easily have conceived the problem to be otherwise; one of moving through the air in desired directions and engaging the water "wind" below (relative current) with scoops. In this case, upwind and downwind would mean directions opposite to the understanding of water animals. This method of attack creates an entirely different vehicle, one with a fixed air centerboard and articulated wet sails.

Actually, the problems are symmetrical and both approaches can be handled with the same set of equations, as was pointed out some time ago by

H. M. Barkla of the Univ. of St. Andrews. It is simply a problem of relative shear between two fluids at their interface. This concept suggests the shape of the ultimate sailing machine: *a platform consisting of a vertical wing in the air joined to an inverted one in the water, all assumed to be stable, buoyant, and deformable for all points of sailing* (see F-25).

To proceed in more detail, certain conditions must be stipulated. For example, before specifying the relative size of the two foils, the designer must select the relative speed desired in the fluids. For the simplest case, equal speeds can be chosen, satisfied approximately by making the areas of the wings inversely proportionate to the fluid densities, about 800:1 at sea level. (It would be roaming afield to take into account Reynolds numbers, Froude numbers and other scaling factors for modeling velocity, viscosity, and scale with more refinement. Density, the overriding determinant, suffices for present purposes.) Thus, if the wings have the same aspect ratio and similar dynamic properties, the span of the airfoil will be about 28 times that of the hydrofoil. The F-25 model shows immediately that, by comparison, ordinary sailboats have far too much immersion in the water and cannot possibly achieve equality of speeds in both fluids. Speed in the water therefore falls far short of theoretical.

It can be shown that equal speed in both fluids represents the optimum when wind and way are nearly perpendicular (F-26). Just as excessive area in the water reduces water speed (and air speed), excessive area exposed to the air reduces air speed (and water speed). Thus, for ratios other than 800:1, you could say that either the air or the water animal has exposed too small a sail.

Only two ways are known for achieving such large ratios. In the most direct, the airfoil becomes a suitably shaped gas-filled bag that supports all but the indispensable underwater components. With lift supplied mostly through air buoyancy, this vehicle partakes of all the load-carrying potential of the displacement hull without its speed limitations; speed is virtually proportional to wind speed over the whole wind spectrum regardless of vehicle scale, whereas payload advances more nearly as the third power of the scale, similar to the hull. The second approach, in which a water-borne body supplies the initial support, cannot of itself meet the ratio; it must be supplanted by true dynamic lift to reduce the immersion of submerged surfaces.

Resorting to hydrofoils may provide lift, but they become an impediment in low winds unless removed. Lift derived from airfoils suffers less penalty. As load increases for a hydrofoil-lifted craft,

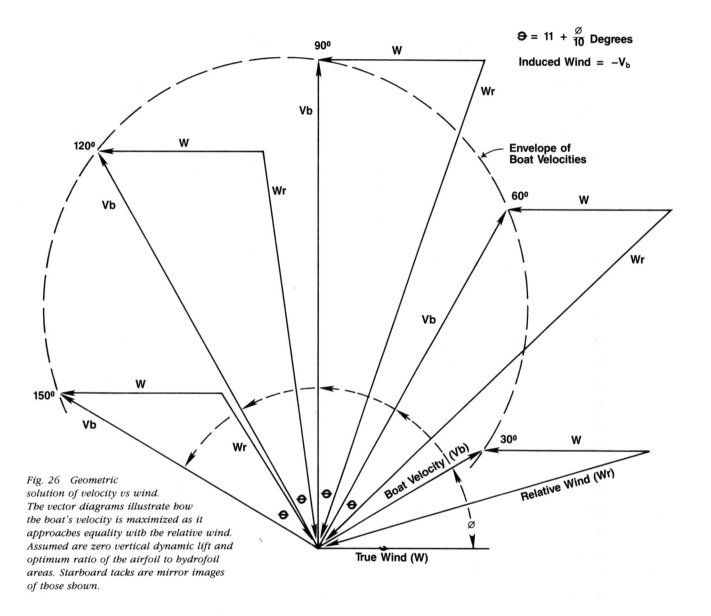

$\theta = 11 + \frac{\emptyset}{10}$ Degrees

Induced Wind $= -V_b$

90°

W

Wr

Vb

120°

W

Wr

Envelope of
Boat Velocities

60°

W

Wr

Vb

Vb

150°

W

Vb

Wr

30°

W

Boat Velocity (Vb)

Relative Wind (Wr)

$\emptyset$

Fig. 26 Geometric
solution of velocity vs wind.
The vector diagrams illustrate how
the boat's velocity is maximized as it
approaches equality with the relative wind.
Assumed are zero vertical dynamic lift and
optimum ratio of the airfoil to hydrofoil
areas. Starboard tacks are mirror images
of those shown.

True Wind (W)

so does the need for higher winds to reach critical lift-off speeds. Until the wind reaches a certain strength, the performance of a sailboat dependent upon hydrodynamic lift does not better an ordinary sailboat's. Unlike the blimp sailboat, the load potential of a dynamically lifted sailboat, much like an airplane, tends to increase only as the square of the scale.

Closeness of the boat's path to the wind (shear line of the fluids), ordinarily understood as "beating," also reveals the nature of the ultimate sailboat. Beating is of extreme importance to a very fast sailboat because of the magnitude of the induced wind produced by the boat's own motion. For example, a sailboat that moves as fast as the true wind only when the apparent wind on board is at least 45° off the bow cannot beat upwind at that speed. It can only do so at a lower speed. The minimum angle of the wind constitutes a performance limit; the closer a boat can sail to the wind's line, the more options it has in direction of motion as well as speed.

In aerodynamic terms (which are not always appropriate to sailing analysis), the L/D ratios of the foils in the two fluids limit the angle off the wind. The highest L/D's are invariably associated with small angles of attack, between 3° and 5°. Sailing upwind emphasizes the smallness of the total angle between the resultant force developed by a foil and its direction of motion in the fluid. This criterion conveniently combines all the factors into one easily visualized number, a force angle, as displayed in F-25, and dispenses with such terms as lift and drag when nothing is being lifted and when, on certain points of sailing, drag contributes to forward motion (as with wind abaft the beam).

The force angle for a very-high-performance wing — a soaring machine, for example — could be as little as 92°, but the enormous aspect ratio required for such a wing imposes upon a sailboat, among other disadvantages, a prohibitive listing moment. Moreover, it is highly questionable that such a force angle could be achieved for the wing in the water, certainly not at high speeds. A more reasonable force angle for each wing, allowing for list and other drag losses, would be closer to 97°, giving 14° for the minimum included angle between the velocity vector in the water and the velocity vector in air. Described in this way, the angle has meaning only when measured on the craft, or when measured as the angle of intersection for a crossed wind and water tunnel in which the craft is stationary by virtue of the dynamic forces acting upon it. More to the point, such a model placed in a cross wind could move at better than three times the wind velocity, regardless of which animal, air or water, took the measurements. Its performance would match an iceboat's.

Now we must choose a parochial point of view, either the air animal's or the water animal's; otherwise we face a possible exercise in confusion when attempting to determine the velocity vectors of interest. We will think primarily as surface animals, but occasionally use the air animal's point of view. Henceforth, platform velocity will mean velocity in the water and wind velocity, of course, will mean velocity through the air, differentiating between true wind (with respect to the water frame) and relative wind (as experienced on the platform).

Relative wind, alternatively called "apparent wind," exceeds true wind except when attempting to sail downwind too close to the true wind's line. The induced wind produced by the boat's own motion through the air must be added vectorially to the true wind to obtain the relative wind. A very fast sailboat would always be sailed to produce a relative wind considerably higher than the true wind, even when the path to its landfall is not direct. In fact, it may be said that to reach destinations around the compass in the least time, a very-high-speed sailboat will be moving about half the time in some direction other than directly toward its destination.

Putting it all together and applying some vector analysis yields the velocity envelope (F-26) for an ideal Sailloon on various points of sailing in a unit true wind. These are limits based on assumptions that the platform is free of parasitic drag and that all foils are uniformly well behaved at all speeds. The realities of course would be somewhat different; they are taken into account in Appendix B.

Does the real thing match theory?

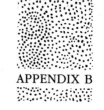
Contribution of Hydrodynamic Lift to Sailing Speed

*G*ood hydrofoil support for Aerohydrofoils, Fliptackers, or other sailboats should reduce wetted area and thereby increase the ratio of airfoil area to hydrofoil area. To determine, in what follows, the quantitative effect of hydrofoil action in the simplest way, lift derived from inclined airfoils will not be taken into account, and the lift of the hydrofoils will be assumed equal to the drift resistance. Accordingly, airfoils will be vertical and hydrofoils inclined at 45° to the horizontal. Also, the angle between airfoil and hydrofoil chords will be fixed at 8°.

To undertake analysis for an entire spectrum of wind strengths would be far too laborious for hand computation. Instead, a wind velocity will be selected to satisfy a desirable condition: i.e., when wind and boat speeds are equal at an arbitrarily small angle off the wind. The initial conditions drawn from preliminary computations will be as described on the next page.

Angles of Attack

Airfoil $\alpha_a = 5°$

Hydrofoil $\alpha_h = 4°$

The difference in angle takes into account two conditions: initial wetted area will be larger than needed for drift resistance and, consequently, the hydrofoil will operate at a smaller drift angle. This will vary as speed and depth of immersion change, as will airfoil angle of attack as angle off the wind increases.

Velocity Vectors

Angle (ϕ) between boat velocity (V_b) and true wind (W) = 34°

Ratio of relative wind (W_r) to W when $V_b = W$ is 1:91

Angle (θ) between V_b and $W_r = 17°$

Area ratio at this point, 200

Foil Lift Equivalents

The lift of the foils can be expressed as

$$F_L = C_L Q A_h V_h^2 \qquad (1)$$

where:

F_L = lift force = load

(M) = anti-drift force, lb

C_L = lift coefficient (= .4 at best L/D)

Q = dynamic constant for water = 1 slug/ft³

A_h = *projected* area of hydrofoils, ft²

It is evident that the immersion (projected) area A_h of the foils depends on the load as well as on V_b. To determine the particular V_b desired, the designer must assume some constancy in the ratio of the load (M) to the area of either the airfoil or the hydrofoil. Since the hydrofoil area is the dependent variable, the constant ratio to establish is obviously M/A_a, in which A_a is the area of the airfoil. Such an assumption of constancy is not completely justified on a strict geometric basis.

As the size of the boat increases, the wind's overturning moment varies as the third power of the size (airfoil area multiplied by moment arm), and the restoring moment generated by the ballast varies as the fourth power (initial hydrofoil volume multiplied by moment arm). The difference in powers permits a two-thirds power increase in airfoil area and a one-third power increase in moment arm. The permissible advance in powers as size increases is therefore 2 2/3 for the airfoil and 3 for the displacement, or load.

This is not quite sufficient to support the position of constancy in the ratio of airfoil area to weight, but other changes that come with size help to make up for the deficiency. For example, taller airfoils have proportionately less root-area in the turbulent air near the water surface, thereby making the lower part more efficient. Also, the higher wind

Further thought models.

speeds near the upper part give an effectively larger angle of attack, which in one way or another can be converted into a greater forward component. Both effects allow an increase in driving force with little increase in overturning moment.

Critical Wind-Speed

The models discussed here easily achieved a ratio of one pound of boat per square foot of airfoil. This is not easily attained for other than very large man-carriers, although with the help of a helium-filled airfoil it can be satisfied at a reduced scale. Numerically, then:

$$F_L = M = A_a \tag{2}$$

Recalling that —

$$A_a/A_h = 200 \tag{3}$$

Substituting in Eq. (1) gives —

$$V_b = \sqrt{500} = 22.4 \text{ ft/s} = 13.4 \text{ kt}$$

This is the speed at which the lifting Aerohydrofoil will sail as fast as the wind at 34° off the wind.

Actually, the situation is a little better than described. The analysis has not taken into account the displacement of the foils, which would have given equality of speeds at a lower wind speed. For the given conditions, however, vertical lift may be considered free of buoyant support and largely dynamic.

Velocity Proportional to Square of Wind Speed

When lift is almost entirely dynamic the supporting force per *unit area* is proportional to V^2; but the immersion of surface-piercing foils automatically adjusts itself to a fixed load by changing height above water and therefore changing the *area* remaining underwater. The adjustment (see *The 40-Knot Sailboat*) effectively brings the drag down to the first power of V_b. Hence, for all other points of sailing:

$$V_b = \left(\frac{W_r}{1.91}\right)^2 \frac{\alpha_a}{5} = -W_b \qquad (4)$$

where W_b is the wind induced by the boat's motion. However, Eq. (4) is valid only for a true wind speed of 13.4 kt. This tells us that the preferred sailing tactic will achieve high relative wind speeds rather than high airfoil angles of attack.

But the angle of attack of the hydrofoil cannot remain constant when the boat sails farther off the wind than 34°. Some increase in drift must be sustained so long as greater speed is derived from an increase in side force that results in reducing the hydrofoil immersion. It is advantageous to accept the higher drift rates that accompany the higher speeds,

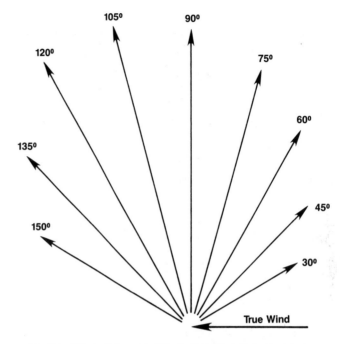

Fig. 27 *Effect of dynamic lift upon performance. Port tacks form a mirror image of the starboard tacks illustrated. Speed on most points of sailing is higher than the wind itself, and at 90 deg to the wind more than twice as much. At 45 and 135 deg the respective upwind and downwind components of velocity signify an ability to beat the wind upwind or downwind in winds above 13.4 kt.*

rather than to project more foil in the water and reduce the speed on these points of sailing. To allow for this we add the following constraint:

$$\frac{\alpha_h}{4} = \left(\frac{W_r}{1.91}\right)^2 \qquad (5)$$

This imposes hydrofoil angles of attack proportional to the side forces on the airfoil.

One could reason that Eq. (5) would also depend on foil immersion, which in turn depends on V_b, but this would not be entirely true. Increases in boat speed as a consequence of increasing the airfoil angle of attack are not encumbered by an equivalent increase in side force, because the airfoil's resultant gains a greater forward component. Thus, at very high angles off the wind, W_r can be relatively low; yet, because airfoil angle of attack can be very high under these conditions, V_b can be high, the foil immersion small, and the drift low.

The angle θ will therefore be the sum of the included angle between foils and the two angles of attack (α_h and α_a), neither of which will very linearly: increasing and decreasing in proportion to W_r^2 and α_a increasing slowly until maximum W_r is reached, whereupon rate gets accelerated. These conditions make it difficult to derive any simple mathematical solution. Iterative processes must now

Table 1   Tabulation of Velocities

ϕ, deg	α_a, deg	α, deg	θ, deg	W_r, kt	V_b, kt	Progress, kt	
34	5.0	4.0	17.0	25.6	13.4	11.1	
45	5.2	5.4	18.6	29.7	18.7	13.2	up-
60	5.4	6.9	20.3	33.5	24.7	12.4	wind
75	5.8	7.8	21.6	35.1	29.2	7.6	
90	6.4	7.9	22.3	35.4	32.7	0.0	
105	7.5	7.0	22.5	33.8	34.7	9.0	
120	9.6	5.4	23.0	29.5	34.0	17.0	down-
135	17.0	4.0	29.0	19.6	26.5	18.8	wind
140	23.0[a]	3.0	34.0	15.4	23.0	17.6	

[a] Slotted airfoil.

be used, wherein values of α_h and α_a are tested for consistency with Eqs. (4) and (5), and with those below, which come from trigonometric relationships:

$$\frac{W_b}{W} = \frac{\text{Sin}(\phi - \theta)}{\text{Sin } \theta} \qquad (6)$$

$$\frac{W_r}{W} = \frac{\text{Sin}(180 - \phi)}{\text{Sin } \theta} \qquad (7)$$

Values computed to two significant figures are given in Table 1 for assumed wind speed of 13.4 kt. F-27 presents a polar diagram of V_b vs ϕ.

A compromise more favorable to upwind sailing could be achieved by curving the hydrofoils so that the lower ends have a higher inclination than the roots. In this way the ratio of the vertical to the horizontal components can be adjusted to fit changing conditions. More lift and more speed would be available in low winds, where sufficient immersion for resisting leeway need not be a problem, and more resistance to drift would be available in high winds, where the craft would be high enough out of the water to be in the desired speed regime.

Other aspects of this computation are worth describing. If the speed of the wind is less than 13.4 kt, the boat will be deeper in the water and will develop more drag. This means it will be slower than the wind at 34° off the wind. On the other hand, the direction of W_r will be more favorable to closer pointing, if desired. If the wind speed is more than 13.4 kt, the boat will be higher and resist leeway less, which means it will not sail at 34° off the wind when moving at the speed of the wind. However, in either case there are compensations: When sailing slower it can point up higher, and if it must point off more, it will make more speed.

For the critical wind speed under discussion, the best average progress against the wind will still be made at about 45° off the true wind. This angle decreases for lower winds and increases for higher winds.

The angle for best progress downwind is not so obvious. It lies somewhere between 125° and 145° off the wind, depending to a large extent on whether the airfoil can be rotated with respect to the hydrofoil, and whether slots and flaps are added to the airfoil for use at high angles of attack. This analysis assumed a fixed relationship for the foils. Although this assumption led to no unreasonable attack angles in the area of interest, at about 140° off the wind the conditions changed sufficiently to warrant adding a slot. If adding the complexity of a slot were acceptable for the design, it could be employed at smaller angles, in which case the best downwind point of sailing could be close to 130° off the wind.

In a very general way, one can infer that a mirror image of sailing to windward exists in the downwind case. When the wind is low, more downwind progress can be made by aligning the boat's direction more nearly with the wind's direction; when the wind is high, better progress is made in a direction farther from the wind's line.

Finally, for downwind components speed can be greater than the wind itself. Upwind components are not quite as good. At best the designer may equal the wind's speed, but this still exceeds the performance of any other kind of sailboat.

The shape of the performance curve for the lifting Aerohydrofoil will vary with changes in wind speed. But since the speed of the lifting Aerohydrofoil is proportional to the square of the wind's speed, the ratio of maximum speed to true wind speed should increase as the wind increases in strength. Thus, if V_b maximum is 35 knots for a value of W equal to 13.4 kt, the boat's maximum speed at higher wind speeds should be as follows:

$$V_{max} = \left(\frac{W}{13.4}\right)^2$$

For a wind of 15 kt, a maximum speed of better than 43 kt should be attainable. Considering that a 15-kt wind is usually accompanied by fairly moderate waves, the surface conditions needed for such speed would not be inconsistent with the required wind.

By comparison with these computations, how well did the closest actual craft do? On a near reach in a wind of 12 kt, the man-carrying Aerohydrofoil attained a speed of 20 kt. Motion pictures of the event indicated an area ratio of about 100. The 9-ft Sailloon equaled that on a beam reach in an 8-kt wind and reached an area ratio of 200. The differences are attributable primarily to the different ratios of payload to foil area, but the results are not far from theory.

Sailing Practice

*J*n the oldest sailing maneuver, drifting downwind, the rate may be increased by increasing
the drag in the air or reducing the drag in the water
or both, but it can never equal the wind speed. To
do so would mean driving the boat downwind with
no force on the sail and no resistance in the water,
both obvious impossibilities.

Sailing at an angle to the wind requires more
sophistication than a windcup and a pointed float.
Certainly no progress can be made upwind without
a true sail and keel. These bodies exhibit the
remarkable property of generating forces nearly perpendicular to the flow of fluids in which they are
immersed. By arranging a suitable angle between sail
and keel, forward motion can be generated around
the compass for conventional sailboats, as shown
in F-28.

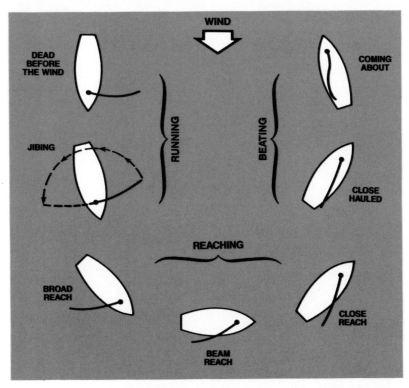

WIND

DEAD BEFORE THE WIND

COMING ABOUT

JIBING

RUNNING

BEATING

CLOSE HAULED

REACHING

BROAD REACH

CLOSE REACH

BEAM REACH

Fig. 28   Conventional sailing practice. The sailing points have meaning only with respect to the true wind. If the apparent wind experienced on the boat is used as a reference, the designations will be in error. For example, a fast boat moving around the points of sailing identified as "reaching" would feel little change in the direction of the apparent wind — not over 30 deg — yet the boat's orientation to the true wind might have changed by more than 90 deg, from downwind to upwind.

The common procedure for changing tack turns the craft across the wind until the sail fills on the opposite side. The track left in the water is sinusoidal, as indicated in F-29. A very fast sailboat would be making the same kind of track downwind, but the sail would be jibing at each change of tack.

Square-rigged ships did not use these procedures. They "wore about," because attempting to change tack in the manner of vessels fitted with fore-and-aft sails would soon place them "in irons," gathering sternway with the wind filling the sails the wrong way. When beating to windward they turned downwind, making a complete circle to come up on the other tack, as traced in F-29. On the other hand, square sails were ideal for changing tack on downwind legs.

The ancient lateen sail presented a similar problem. It was recognized early on that the lateen sail pulled best when it filled in a smooth curve. The question was how to get the sail from one side of the mast to the other. To accomplish this the crew followed the procedure for square-rigged vessels except that the operation was completed by pulling the sail's loose foot forward and then around the other side of the mast.

Means for coming about are certainly not exhausted with these descriptions. The Aerohydrofoil, in common with the Pacific flying proa, change tack by simply reversing direction. The proa's lateen sail is brought around the mast when the hull is broadside to the wind at which time the rudderman and his paddle change ends (actions eliminated in the Aerohydrofoil design). The big Sailloon of F-8 would be brought about in identically the same manner, producing a saw-toothed track in the water, as traced in F-29.

In the Monomaran the sail and hydrofoils reverse direction to come about whereas the hull proceeds independently, moving in the direction of its bow. The combined motions being almost indescribable, no attempt will be made here to trace its track.

Of course, the Fliptacker has its own way, which matches none of the above; and it has received enough attention.

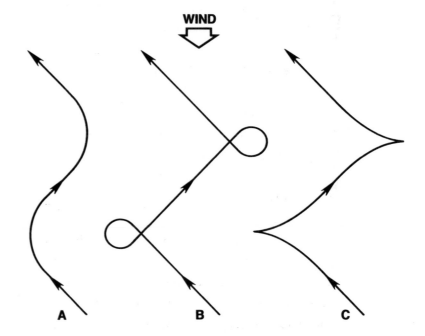

Fig. 29 Beating tacks. Track A depicts the most familiar manuever for coming about, practiced by craft with fore-and-aft sails. Square-rigged ships had to employ track B in order to avoid the danger of sternway. The Pacific "Flying Proa," capable of reversing direction, employed yet another strategy. It turned broadside to the wind, stopped to reverse the sail and rudder, and proceeded in the opposite direction, exactly as does the Aerohydrofoil.

WIND

A   B   C

Origins of the Sailboat

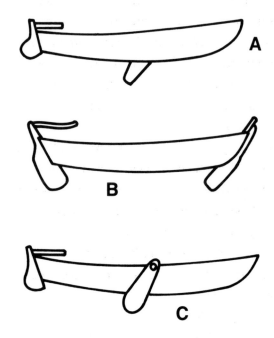

Fig. 30 Movable devices for resisting leeway. Articulating keels, or drop-keels — conveniences invented mainly for shallow-water sailing — also provide a fringe benefit insofar as they can be raised to reduce drag in sailing downwind. Sketches A, B, and C show three symmetrical drop-keel arrangements: common centerboard, Indo-Chinese stemboard, and Dutch leeboards.

*T*he explosive burst of European curiosity and exploration in the Age of Discovery found its most dramatic expression through the sailing ship. Furnished with a true rudder and flexible combinations of square and lateen sails, all held together with a greatly advanced shipwright art, new sailing ships turned the previous barrier of the ocean into a roadstead to all parts of the world.

What other sailing technology confronted intrepid Western Europeans as they left the Mediterranean sphere of influence that extended beyond the North Sea at one end and into the Arabian Sea at the other?

Coming onto the Eastern Asiatic shores the Europeans saw fully battened balanced lug sails controlled with multiple sheet lines and retractable balanced rudders. Were these uniquely different ships of independent origin or were they derived from Western ships by design links that have since disappeared? A case for the common origin of Western and Eastern sails emerges when we examine the properties of the ubiquitous lateen sail, known all about the Arabian peninsula. This ancient descendant of the square sail (tilt the yard of a square sail and remove a corner of its foot and you have a lateen; see F-30) hardly needs more than gradual transitions to trace its westward evolution into the sprit, gunter, lug, leg-o'-mutton, and gaff-and-boom sails. Moving in the other direction, one can visualize the simple lug by fitting a lateen with a boom fastened to the mast. From this version the way to the Chinese lug is straightforward.

The underwater components are not so easily related. The notion of the balanced retractable rudder is more readily stimulated by the balanced lug sail than by steering oars or the rudders existing in other parts of the world. The early Chinese invention of bulkheading to improve the safety of a ship is a similar case in point. Links between these features and those on Western ships do not exist.

A stronger case may be made for a separate origin of sailing devices among the islands of the Pacific. Surely, Polynesia was the home of the multiple-hulled sailboat, a craft that may have radiated into the Indian Ocean at about the time of Magellan's circumnavigation, without influencing ship design on either side of the Eurasian continent.

We know from trade items and other tokens such as gunpowder and the compass that the Eastern and Western sides of the Old World were in communication well before the Age of Discovery. The dearth of information exchange among these regions for hundreds of years in regard to bulkheaded ships, balanced sails, and rudders requiring little manual effort for control, stemboards, leeboards and centerboards, multiple hulls, etcetera remains a mystery.

Some Eastern devices were extremely useful, as I found when I replaced a centerboard and its box in a Snark with a swiveling stemboard. Not only was the central obstruction in this tiny boat eliminated, but also this retractable means of resisting leeway was completely inspectable and jam-free. Why did the only possible other retractable means to avoid the centerboard box — the leeboard — remain parochially in Europe as the stemboard did in Southeast Asia (see F-31,32)?

Hardest to explain: Why the ocean-spanning multiple-hulled craft that sailed circles around Magellan ships never reached the Americas or spread around the coasts of Africa and Australia, both of

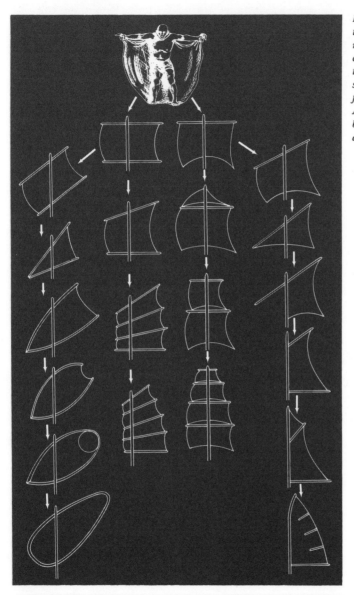

Fig. 31 Sail evolution. Starting with the right, or most easterly column, each suggests the regional progression of sail designs for Oceania, Eastern Asia, and two Western branches of European development.

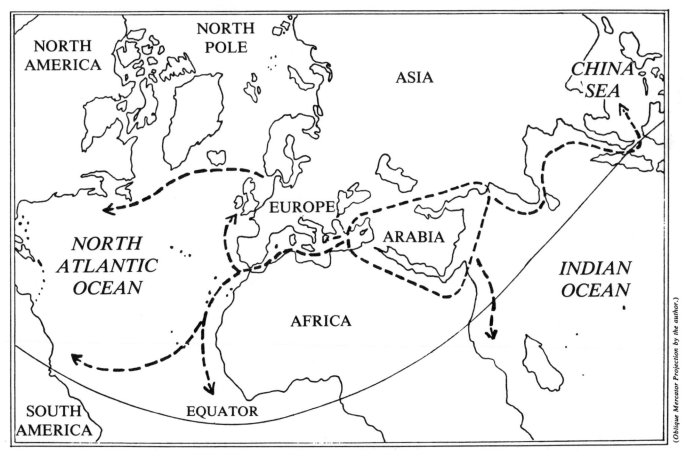

Fig. 32 Migration of the single-hulled boat. The area within the marked triangle is the most probable home of the first single-hulled sailboats. Various records verify the western routes of migration, shown by dashed arrows. The eastern route to the China Seas is suggested by some similarities between boats found along the route and older boats once found in the marked triangle.

(Oblique Mercator Projection by the author.)

which were accessible by island-hopping over distances shorter than those regularly traversed by the Polynesians.

If ever an emerging culture needed the boon of water transport it was the indigenous people of the stretched-out Americas. Yet after witnessing Viking sailboats long before Columbus, the resourceful Eskimos and Indians, who were thoroughly conversant with watercraft, never adopted the sail. From the Arctic to the Antarctic no sailboat was encountered by the first Europeans. Reports of sailing "balsas" along Peru and "zangadas" along Brazil at the time of South American discovery are, in my opinion, highly suspect. Not only are the two locales thousands of miles apart without connecting waterways, but both craft are too improbably identical. Both are rafts with board-like drop keels. On the other hand, if these reports are true, why were sailboats ignored by advanced cultures in Mesoamerica and Peru that badly needed this valuable aid to trade?

Above all, we should ask why Western Europeans were the only ones to employ the ship for the purpose of discovery when ships manned by the Arabs were equally capable and those of the Chinese even more so. Western Europeans discovered the rest of the world, not the reverse. Moreover, where Western Europeans have concentrated, around the North Atlantic, the ship has reached the pinnacle of advancement, as have all other phases of civilized life.

Where ship development in more remote locales has approached the same status, we find a remarkable borrowing of European culture and a concomitant advance in politics, science, literature and the other arts. The history of high civilization seems to be the history of Europe and the primary vehicle of its dissemination seems to have been the wind-driven ship.

Acknowledgments

First among the shop troopers who helped me were Tony Hines of China Lake, CA, and Jimmy "Skeeter" Burns of Dahlgren, VA. These two and I spoke the language of materials and machines. Together we moved fast and economically. Later contributions came from Jim Post and Joe Stacey at Dahlgren, VA; but by far the most advancement was made with Frank Delano of Warsaw, VA, when I was in retirement on Rosier Creek, an estuary of the Potomac River. Frank became at once the Supply Officer, Intelligence Officer, Joint Inventor, Negotiator, Partner, and Test Pilot for about eight years of highly satisfying investigations into the forms of the Ultimate Sailboat. Eventually replacing him was Ralph Gitomer, who still collaborates with me on fabrication and documentation. Together we hope to wind up the story with a flourish.

In recognizing Evan Mann's contribution to this book's graphic clarity I pay tribute not only to his talents but also to the blessings of nepotism; he is my stepson. Should these probings into the Ancient Interface be of value to those who have learned of them, a debt of gratitude is owed John Newbauer, long-time Editor-in-Chief of AIAA's *Astronautics & Aeronautics* magazine, who urged and aided the publication of my findings.

Sources

Barkla, H. M., "The Physics of Sailing," *Bulletin of the Institute of Physics and the Physical Society,* U.K., March 1964. Figure 6 of this paper proposes a design in which overturning moments are cancelled. It appears to be a generic duplicate of the "Sailplane" invented and successfully tested at 19 knots by Malcolm McIntyre three decades earlier.

Casson, Lionel, *Ships and Seamanship in the Ancient World,* Princeton University Press, Princeton, NJ, 1971.

Hadden, A.C., and Hornell, James, *Canoes of Oceania,* Vol. 1, Bernice P. Bishop Museum, Honolulu, 1936.

Hornell, James, *South American Balsas,* Cambridge University Press, Cambridge, UK, 1942.

Marchaj, C. A., *Aero-Hydrodynamics of Sailing,* International Marine Publishing, Camden, Maine, 1988. Of particular interest is the New Zealand craft "Aquarius" (on page 725), which looks like a Fliptacker. Unfortunately, Marchaj gives it short shrift and fails to identify its inventor. Aquarius may represent an instance of simultaneous and independent emergence of similar ideas, regrettably too late for full recognition in this book.

McIntyre, Malcolm, "The Sailplane," *Yachting*, Vol. 55, No. 2, 1934, pp. 62-68. McIntyre's work on high-speed stable sailboats preempted everyone, yet sad to report, receives little acknowledgment from modern professionals seeking the same ends. My own analysis and experiments may properly be considered an extension of his pioneering efforts, which he surely would have pursued had he lived longer.

Smith, Bernard*, The 40-Knot Sailboat,* Grosset & Dunlap, New York, 1963.

"The Aerohydrofoil," Proceedings of the First AIAA Symposium on Aero/Hydronautics of Sailing, Vol. 8, 1969.

"The Monomaran," Proceedings of the Seventh AIAA Symposium on Aero/Hydronautics of Sailing, Vol. 1, 1976.

"New Approaches to Sailing," *Astronautics & Aeronautics,* Vol. 18, No. 3, March 1980.

Index

American Institute of Aeronautics and Astronautics, Inc.
Washington, D.C.

Library of Congress Cataloging in Publication Data

Sailloons and Fliptackers • by Bernard Smith

ISBN 0-930403-65-7

design • Sara Bluestone
drawings • Evan Mann
end-paper motif • Susan Smith
editing • John Newbauer
composition • Simki Michael

type • Garamond
body paper • 80# White Endleaf
dust jacket • NASA pseudocolor space-radar
image of 250-m-long ocean waves

Printed in the United States of America